Solar System Planets

COLORING BOOK

PHARAOHS DESIGNERS

This is a Bleed Through Page If You Are Using a Pen or a Coloring Marker!

Find Our Great Books By searching for **PHARAOHS DESIGNERS** *on Amazon*

PHARAOHS DESIGNERS

MERCURY

This is a Bleed Through Page If You Are Using a Pen or a Coloring Marker!

Find Our Great Books By searching for **PHARAOHS DESIGNERS** *on Amazon*

PHARAOHS DESIGNERS

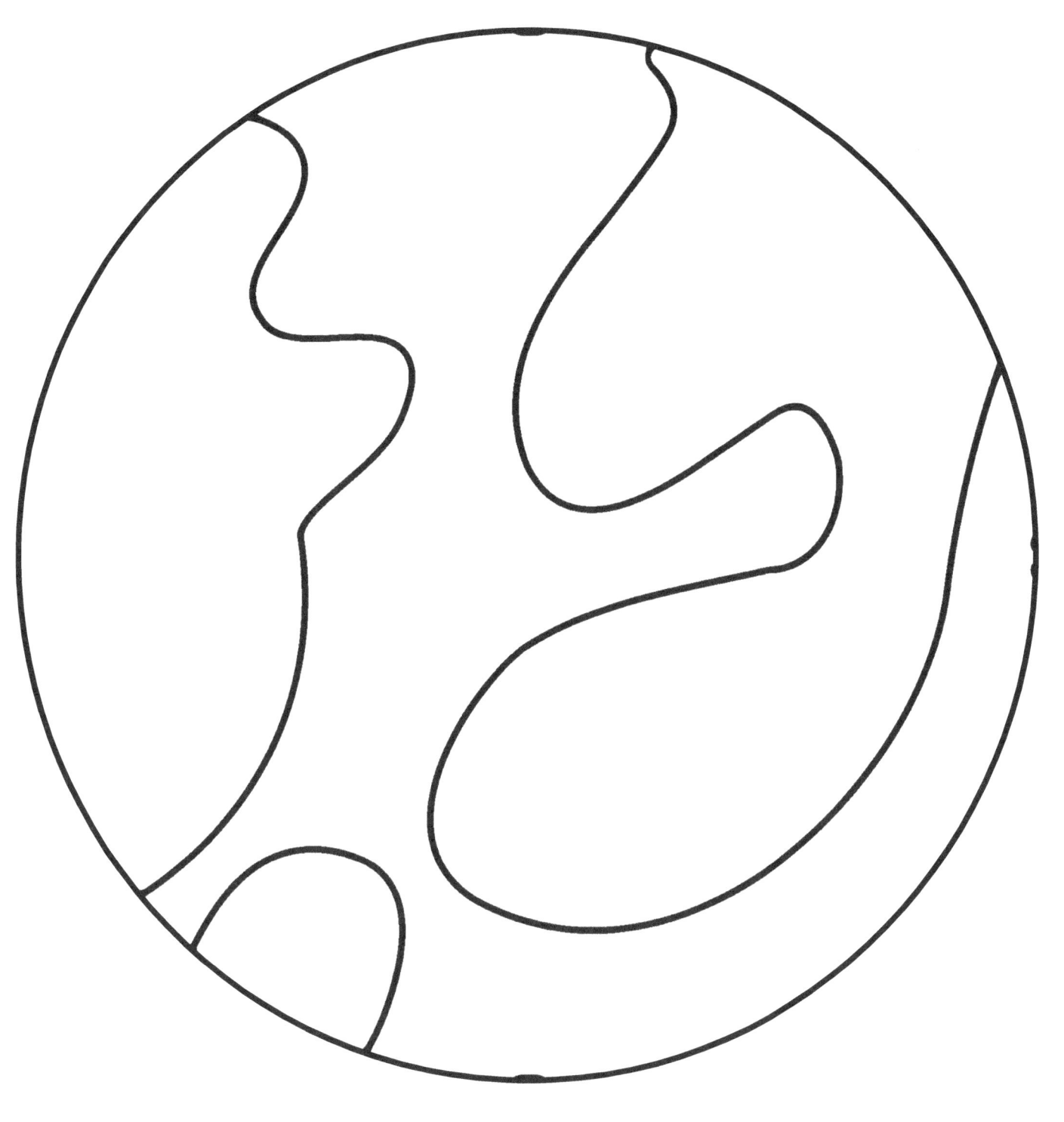

VENUS

This is a Bleed Through Page If You Are Using a Pen or a Coloring Marker!

Find Our Great Books By searching for **PHARAOHS DESIGNERS** *on Amazon*

PHARAOHS DESIGNERS

EARTH

This is a Bleed Through Page If You Are Using a Pen or a Coloring Marker!

Find Our Great Books By searching for **PHARAOHS DESIGNERS** on Amazon

MOON

This is a Bleed Through Page If You Are Using a Pen or a Coloring Marker!

Find Our Great Books By searching for **PHARAOHS DESIGNERS** *on Amazon*

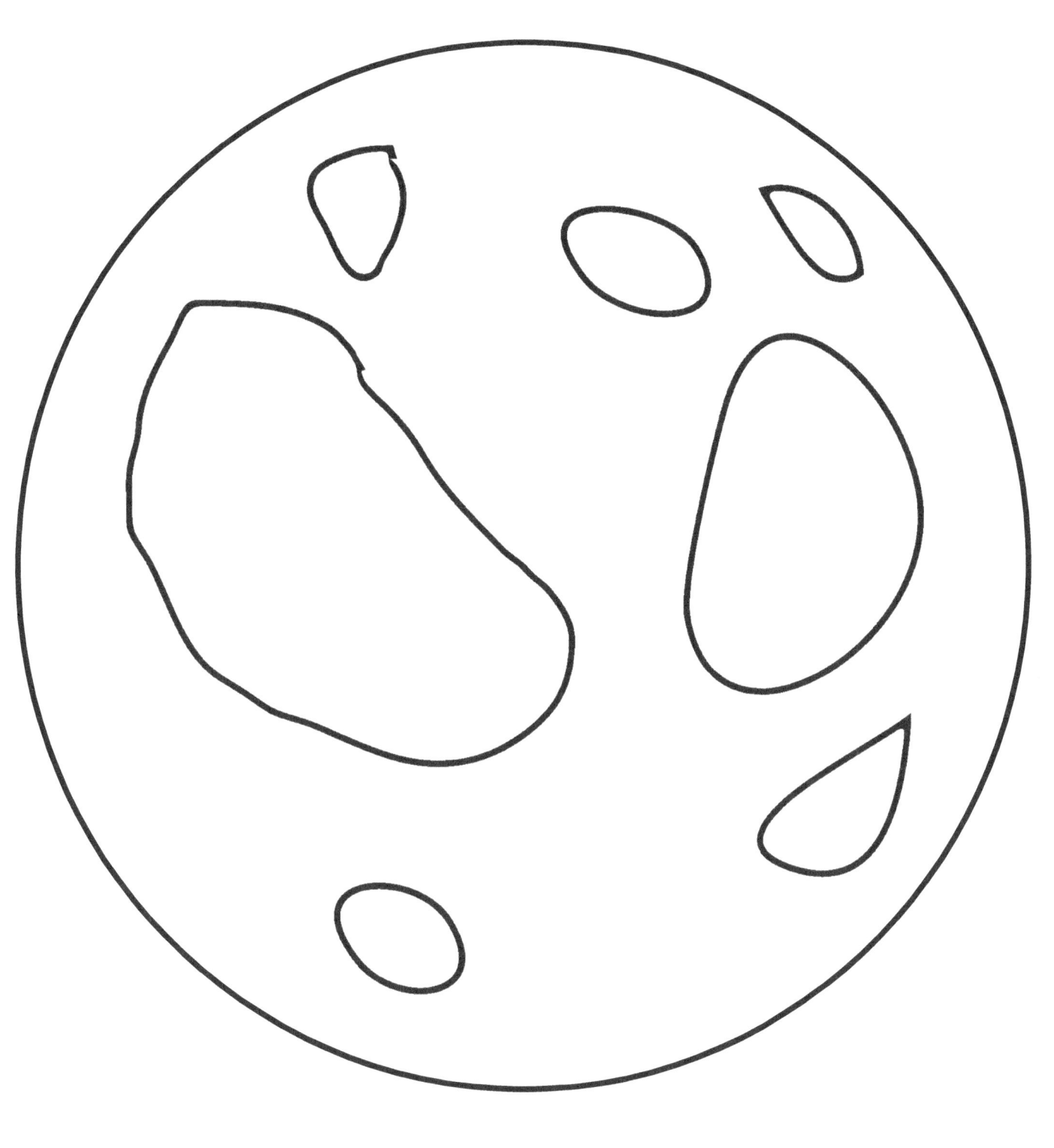

MARS

This is a Bleed Through Page If You Are Using a Pen or a Coloring Marker!

Find Our Great Books By searching for **PHARAOHS DESIGNERS** *on Amazon*

PHARAOHS DESIGNERS

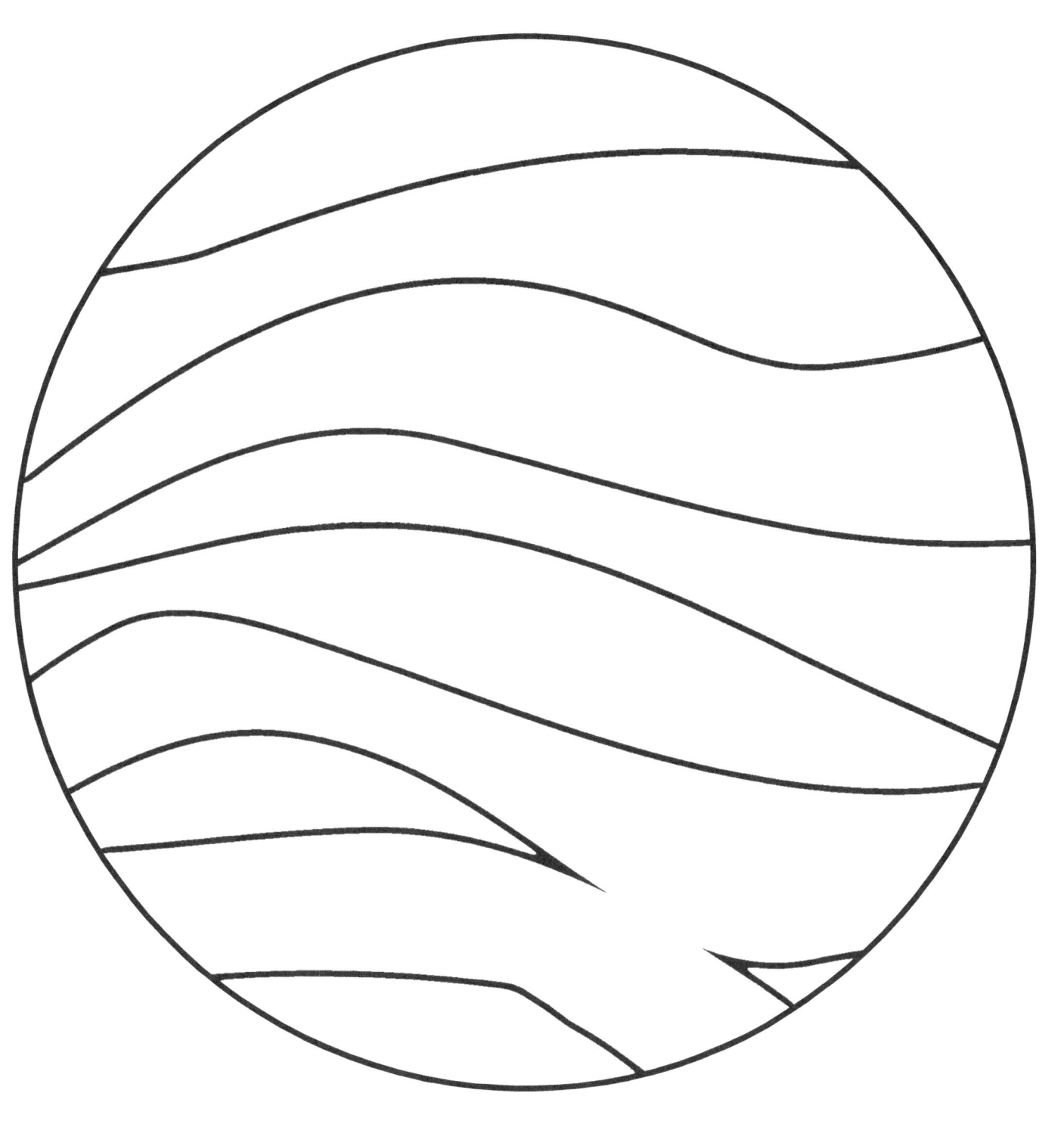

JUPITER

This is a Bleed Through Page If You Are Using a Pen or a Coloring Marker!

Find Our Great Books By searching for **PHARAOHS DESIGNERS** *on Amazon*

PHARAOHS DESIGNERS

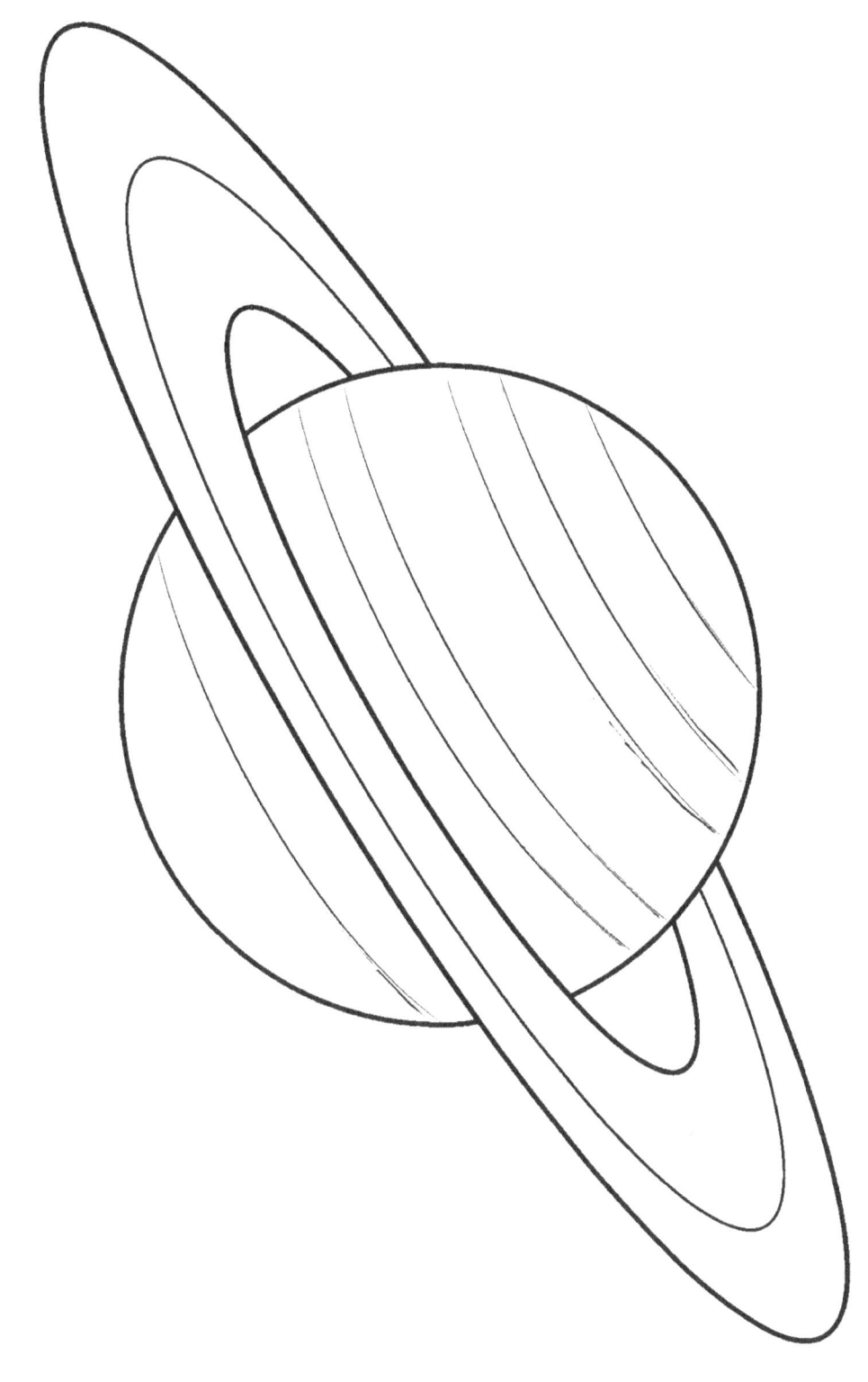

SATURN

This is a Bleed Through Page If You Are Using a Pen or a Coloring Marker!

Find Our Great Books By searching for **PHARAOHS DESIGNERS** *on Amazon*

PHARAOHS DESIGNERS

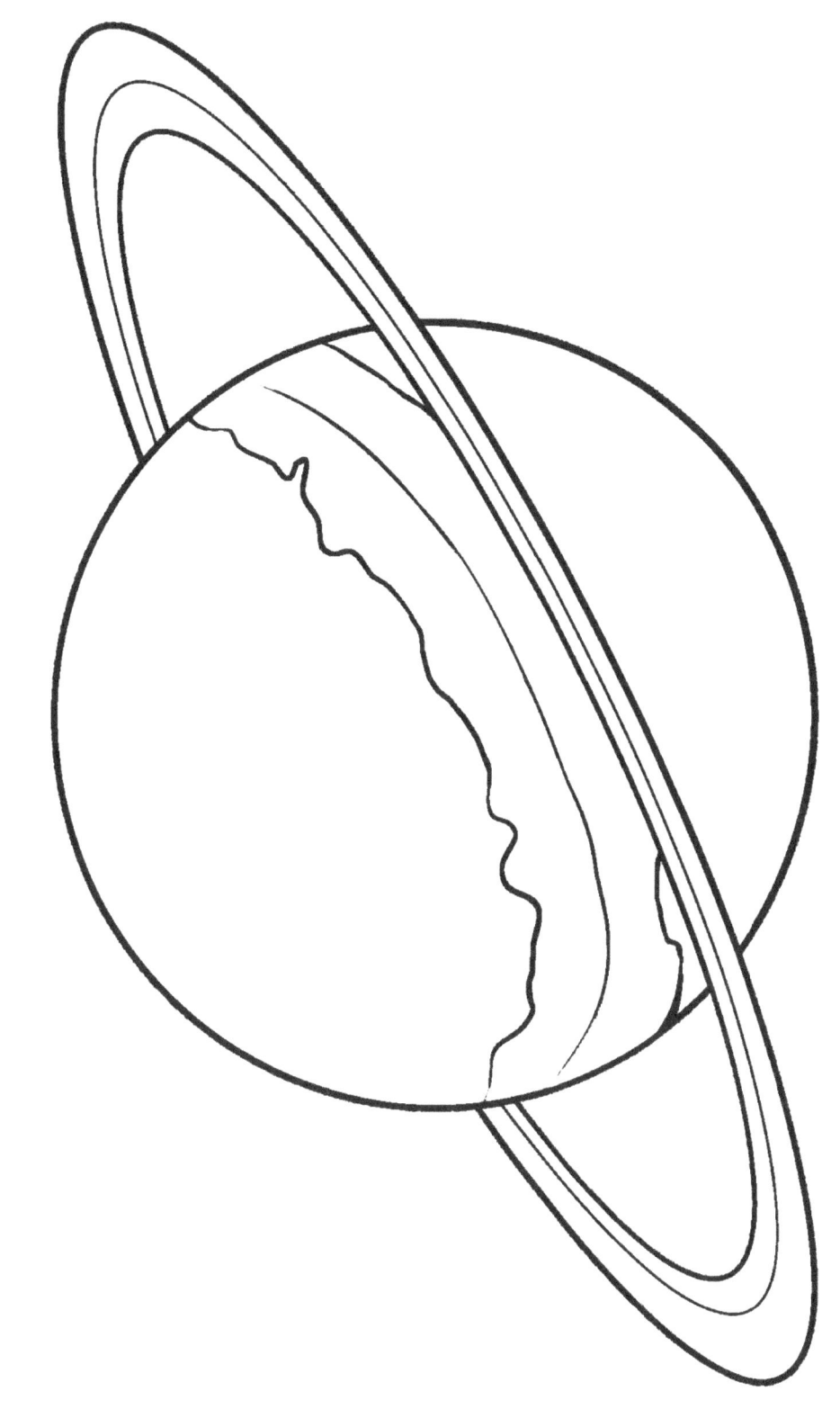

URANUS

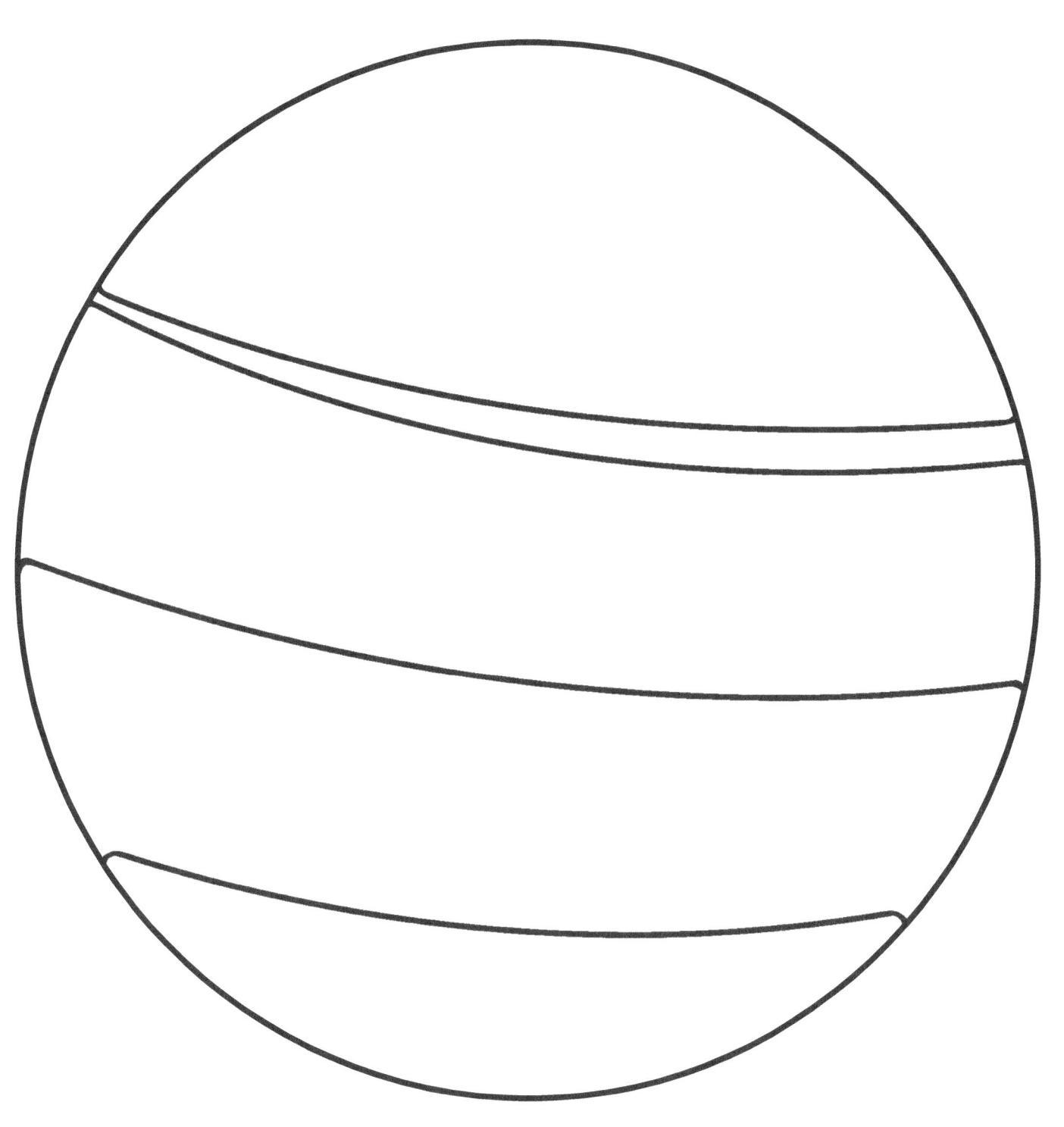

NEPTUNE

This is a Bleed Through Page If You Are Using a Pen or a Coloring Marker!

Find Our Great Books By searching for **PHARAOHS DESIGNERS** *on Amazon*

SUN

This is a Bleed Through Page If You Are Using a Pen or a Coloring Marker!

Find Our Great Books By searching for **PHARAOHS DESIGNERS** *on Amazon*

PHARAOHS DESIGNERS

SOLAR
SYSTEM PLANETS

WRITE MY NAME & COLOR ME

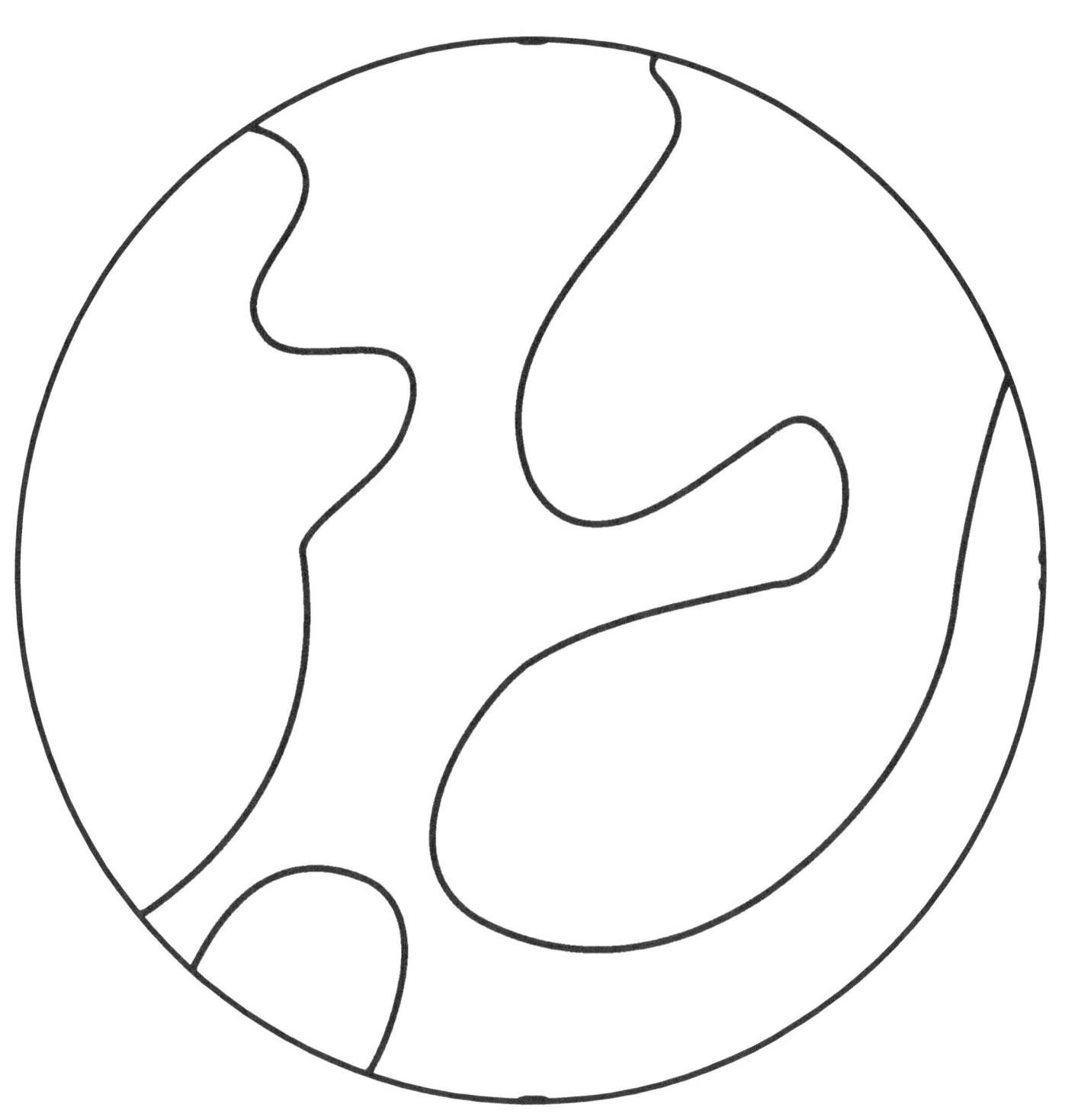

This is a Bleed Through Page If You Are Using a Pen or a Coloring Marker!

Find Our Great Books By searching for **PHARAOHS DESIGNERS** *on Amazon*

PHARAOHS DESIGNERS

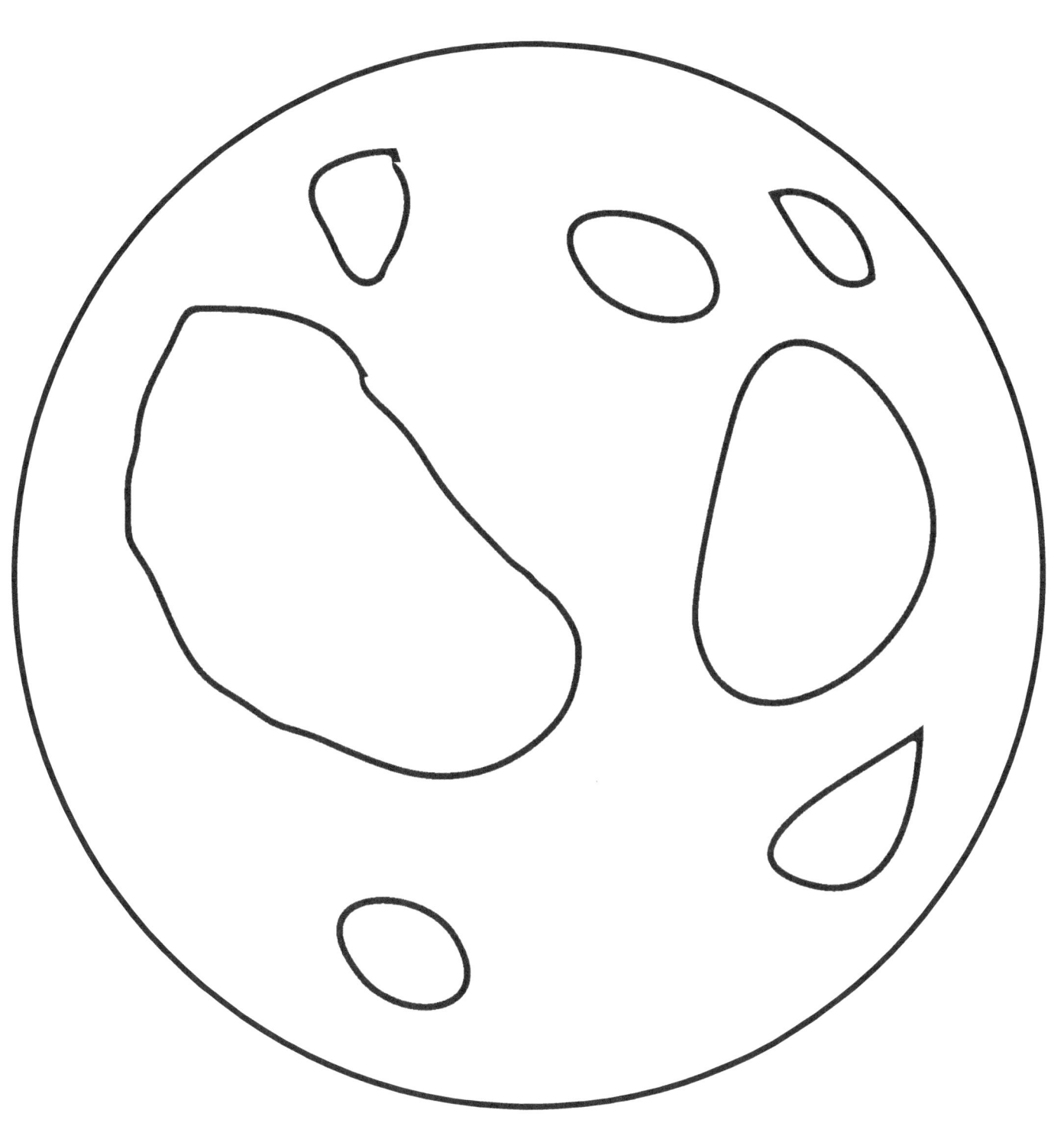

This is a Bleed Through Page If You Are Using a Pen or a Coloring Marker!

Find Our Great Books By searching for **PHARAOHS DESIGNERS** *on Amazon*

PHARAOHS DESIGNERS

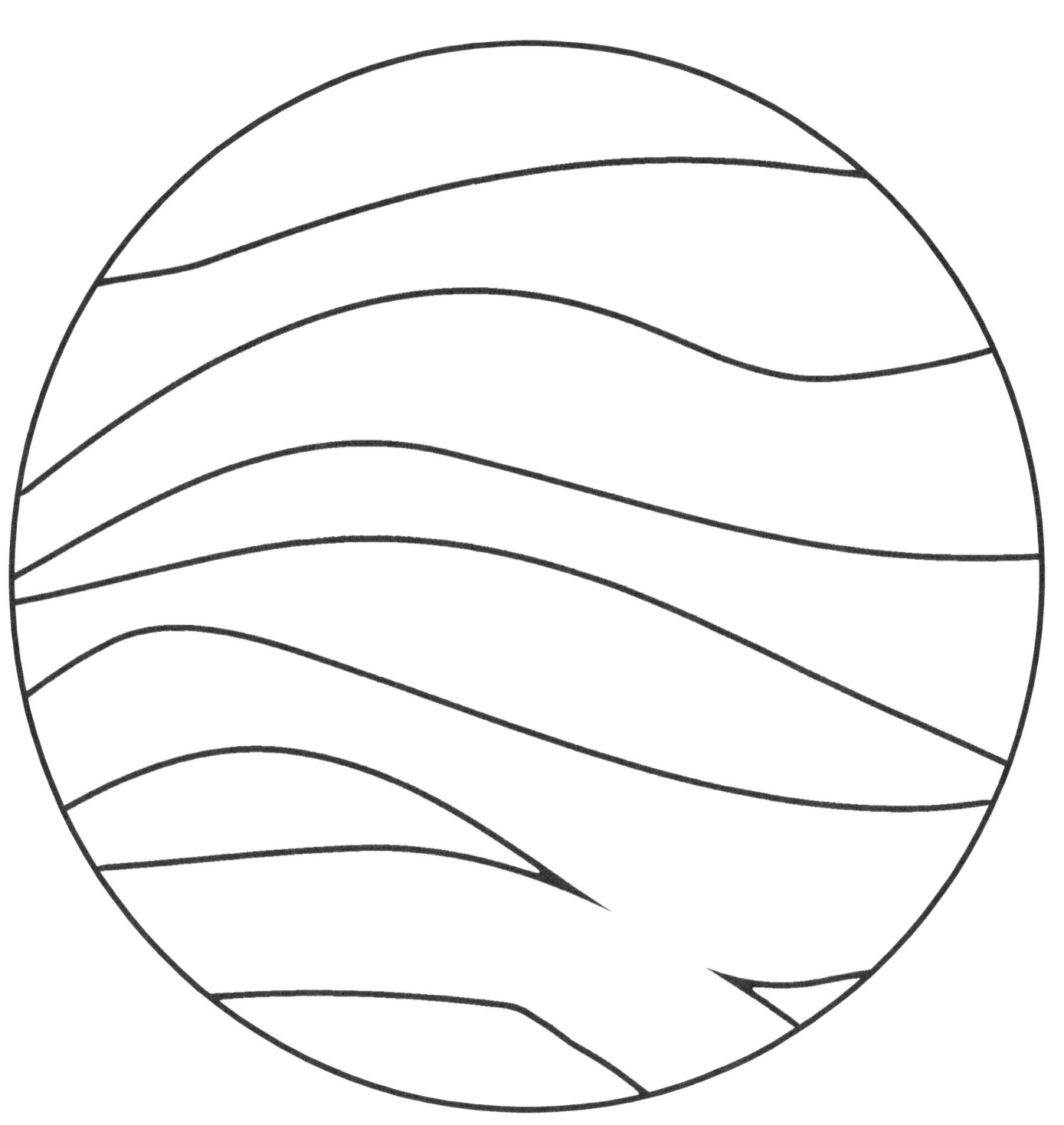

This is a Bleed Through Page If You Are Using a Pen or a Coloring Marker!

Find Our Great Books By searching for **PHARAOHS DESIGNERS** *on Amazon*

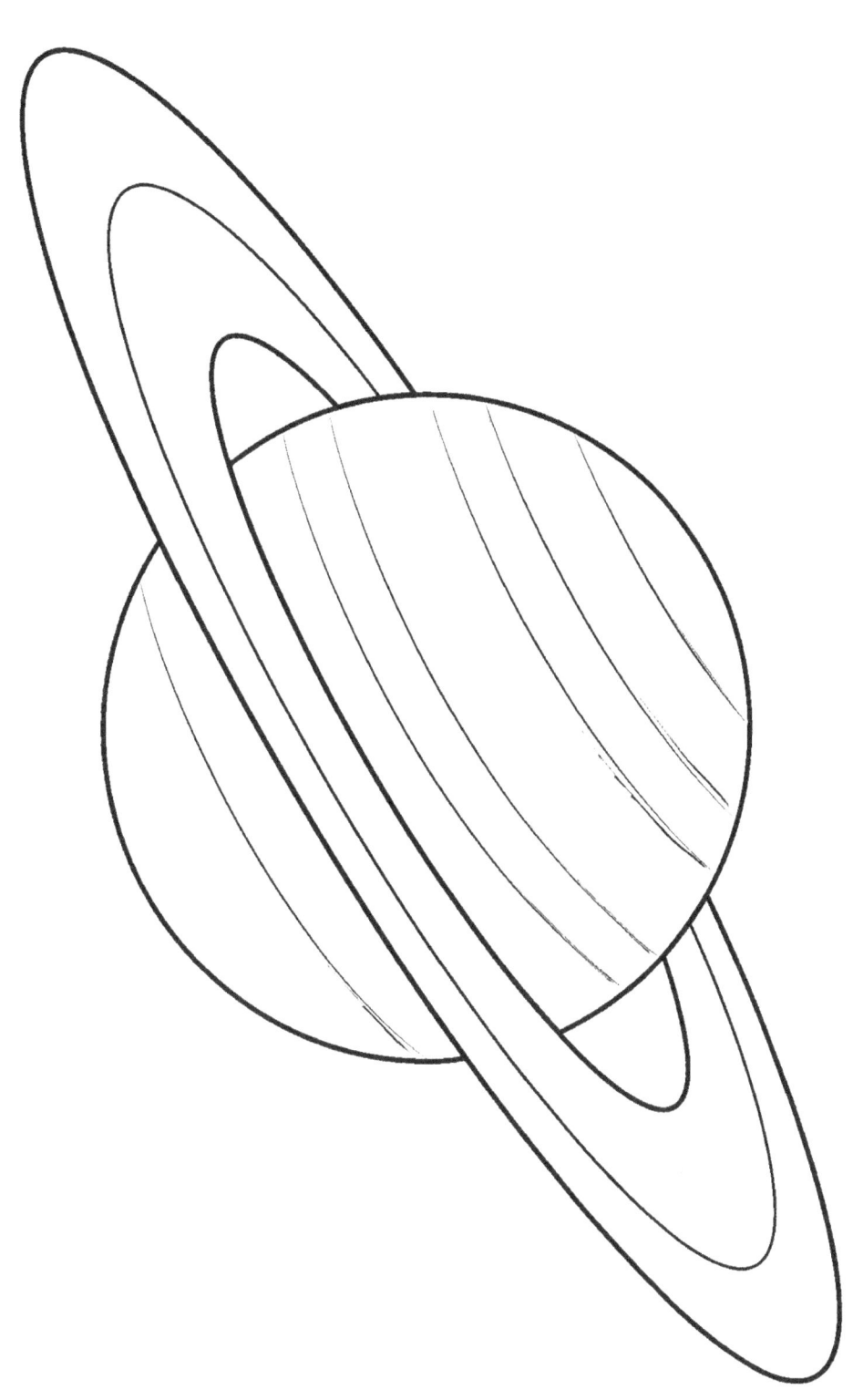

This is a Bleed Through Page If You Are Using a Pen or a Coloring Marker!

Find Our Great Books By searching for **PHARAOHS DESIGNERS** *on Amazon*

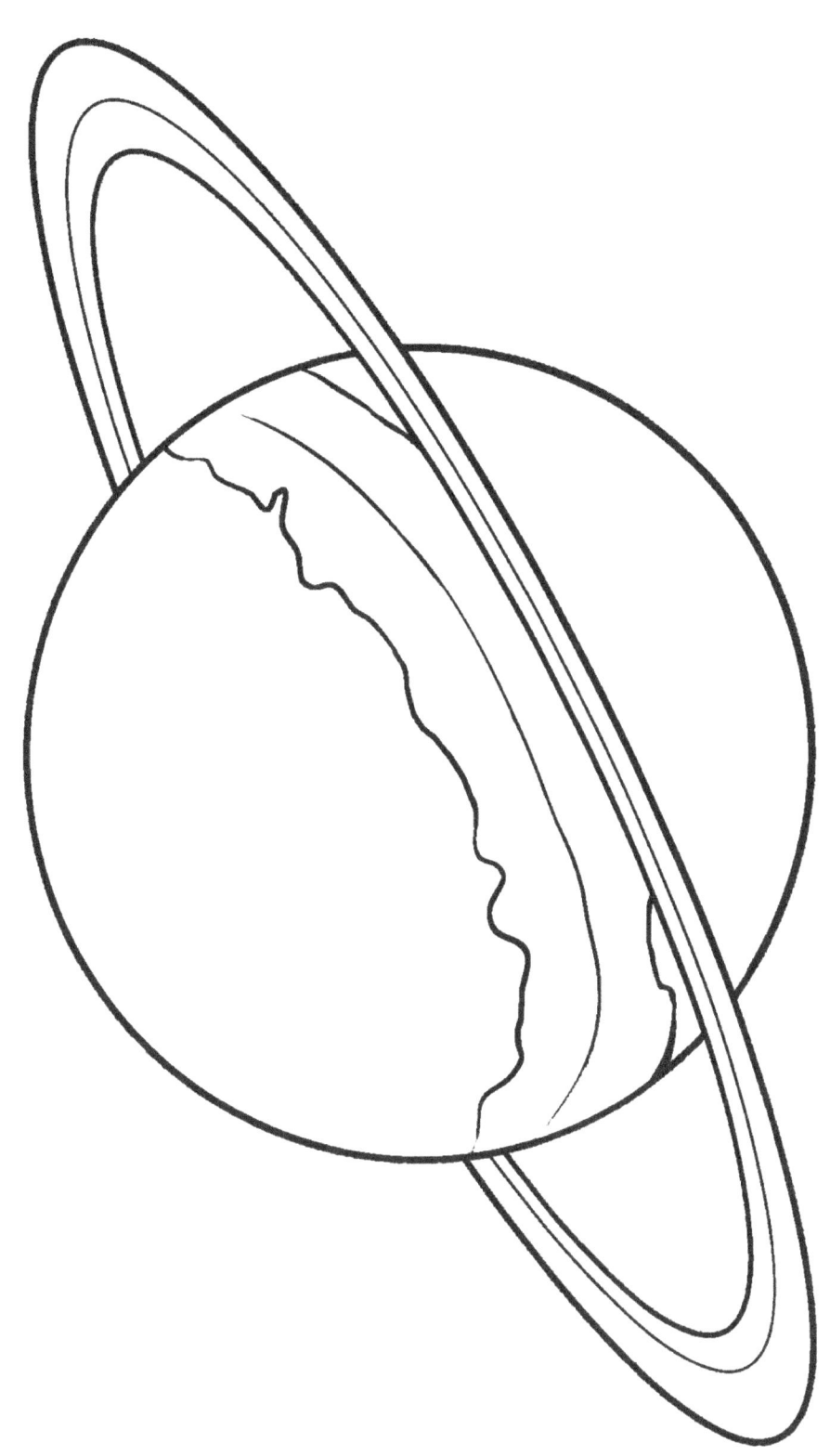

This is a Bleed Through Page If You Are Using a Pen or a Coloring Marker!

Find Our Great Books By searching for **PHARAOHS DESIGNERS** *on Amazon*

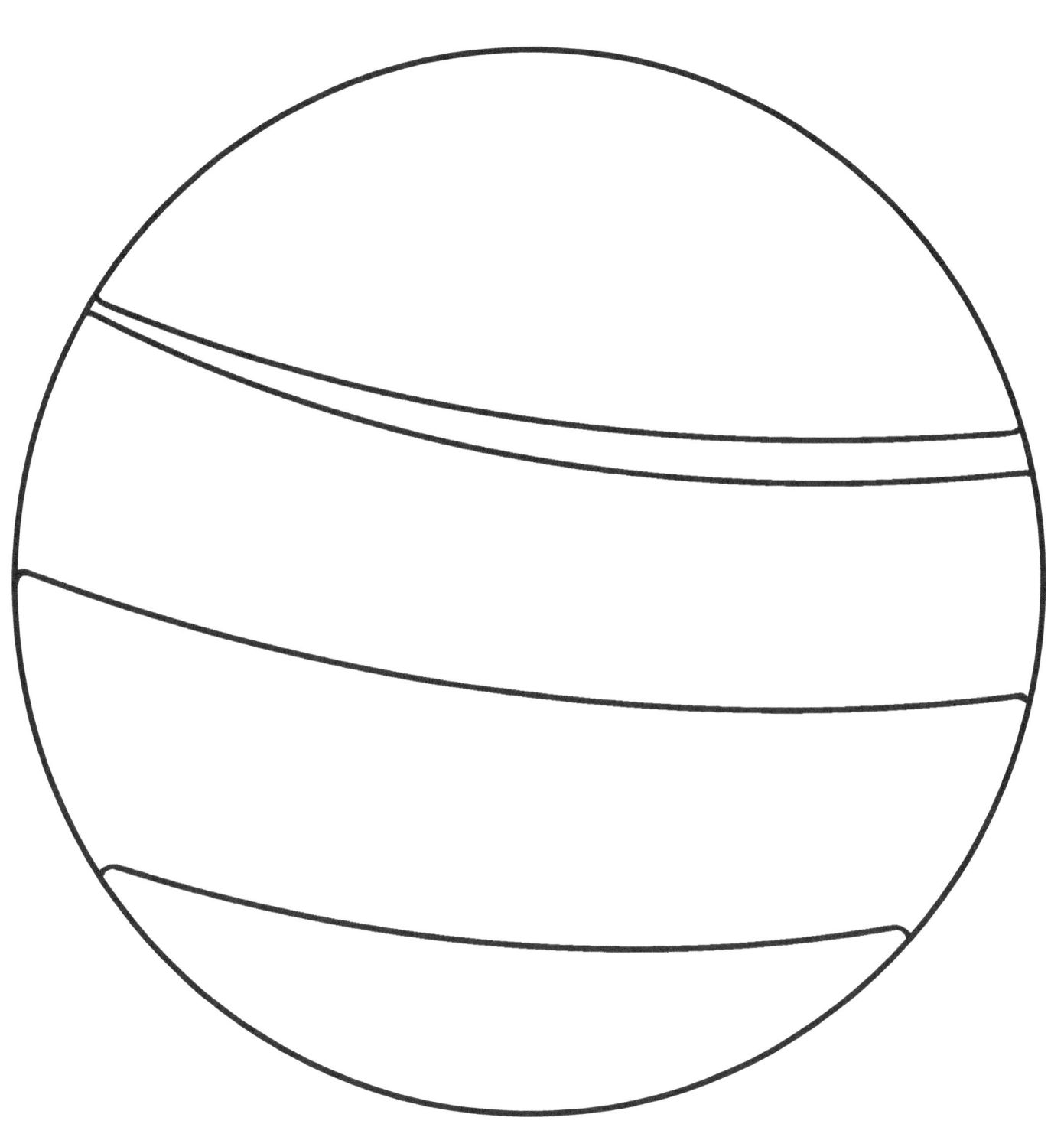

This is a Bleed Through Page If You Are Using a Pen or a Coloring Marker!

Find Our Great Books By searching for **PHARAOHS DESIGNERS** *on Amazon*

This is a Bleed Through Page If You Are Using a Pen or a Coloring Marker!

Find Our Great Books By searching for **PHARAOHS DESIGNERS** *on Amazon*

DRAW & COLOR ME

MERCURY

This is a Bleed Through Page If You Are Using a Pen or a Coloring Marker!

Find Our Great Books By searching for **PHARAOHS DESIGNERS** *on Amazon*

VENUS

EARTH

MOON

MARS

This is a Bleed Through Page If You Are Using a Pen or a Coloring Marker!

Find Our Great Books By searching for **PHARAOHS DESIGNERS** *on Amazon*

PHARAOHS DESIGNERS

JUPiTER

This is a Bleed Through Page If You Are Using a Pen or a Coloring Marker!

Find Our Great Books By searching for **PHARAOHS DESIGNERS** *on Amazon*

PHARAOHS DESIGNERS

SATURN

This is a Bleed Through Page If You Are Using a Pen or a Coloring Marker!

Find Our Great Books By searching for **PHARAOHS DESIGNERS** *on Amazon*

PHARAOHS DESIGNERS

URANUS

This is a Bleed Through Page If You Are Using a Pen or a Coloring Marker!

Find Our Great Books By searching for **PHARAOHS DESIGNERS** *on Amazon*

PHARAOHS DESIGNERS

NEPTUNE

This is a Bleed Through Page If You Are Using a Pen or a Coloring Marker!

Find Our Great Books By searching for **PHARAOHS DESIGNERS** *on Amazon*

This is a Bleed Through Page If You Are Using a Pen or a Coloring Marker!

Find Our Great Books By searching for **PHARAOHS DESIGNERS** *on Amazon*

SOLAR
SYSTEM PLANETS